BEI GRIN MACHT SICH IHR WISSEN BEZAHLT

- Wir veröffentlichen Ihre Hausarbeit, Bachelor- und Masterarbeit

- Ihr eigenes eBook und Buch - weltweit in allen wichtigen Shops

- Verdienen Sie an jedem Verkauf

Jetzt bei www.GRIN.com hochladen und kostenlos publizieren

Bibliografische Information der Deutschen Nationalbibliothek:

Die Deutsche Bibliothek verzeichnet diese Publikation in der Deutschen Nationalbibliografie; detaillierte bibliografische Daten sind im Internet über http://dnb.d-nb.de/ abrufbar.

Dieses Werk sowie alle darin enthaltenen einzelnen Beiträge und Abbildungen sind urheberrechtlich geschützt. Jede Verwertung, die nicht ausdrücklich vom Urheberrechtsschutz zugelassen ist, bedarf der vorherigen Zustimmung des Verlages. Das gilt insbesondere für Vervielfältigungen, Bearbeitungen, Übersetzungen, Mikroverfilmungen, Auswertungen durch Datenbanken und für die Einspeicherung und Verarbeitung in elektronische Systeme. Alle Rechte, auch die des auszugsweisen Nachdrucks, der fotomechanischen Wiedergabe (einschließlich Mikrokopie) sowie der Auswertung durch Datenbanken oder ähnliche Einrichtungen, vorbehalten.

Impressum:

Copyright © 2015 GRIN Verlag, Open Publishing GmbH
Druck und Bindung: Books on Demand GmbH, Norderstedt Germany
ISBN: 9783668255623

Dieses Buch bei GRIN:

http://www.grin.com/de/e-book/335479/synthesis-of-methylated-bucky-bowls-by-hf-elimination

Manuel Langer

Aus der Reihe: e-fellows.net stipendiaten-wissen

e-fellows.net (Hrsg.)

Band 1894

Synthesis of Methylated Bucky Bowls by HF elimination

GRIN Verlag

GRIN - Your knowledge has value

Der GRIN Verlag publiziert seit 1998 wissenschaftliche Arbeiten von Studenten, Hochschullehrern und anderen Akademikern als eBook und gedrucktes Buch. Die Verlagswebsite www.grin.com ist die ideale Plattform zur Veröffentlichung von Hausarbeiten, Abschlussarbeiten, wissenschaftlichen Aufsätzen, Dissertationen und Fachbüchern.

Besuchen Sie uns im Internet:

http://www.grin.com/

http://www.facebook.com/grincom

http://www.twitter.com/grin_com

Table of Contents

Table of Contents 1
Introduction 2
Results and Discussion 3
Summary and Outlook 7
Experimental Section 7
Literature 14

Introduction

For the simulation of complex systems in chemistry or physics the need of very powerful computers grew more and more important over the last decades. As electronic circuits became smaller over the years, more efficient microchips could be fabricated. But the 'classic' electronics are limited and therefore new methods had to be developed, like organic electronics. Here, organic molecules are placed between electrodes in order to perform the basic operations known from digital electronics.[1,2] The basic idea is, to use a optically switchable single molecular core connected to the electrodes on each side.[3] Upon irradiation, most of the applied molecules undergo a conformational change from an open, linear form to a closed ring creating a conjugated π-system throughout the molecule. The π-system enables current flow comparable to a normal wire. The core molecules often consist of dithienylcyclopentenes[4] or difurylcyclopentenes[5,6], coupled to metal electrodes through anchoring groups. The most prominent anchoring groups are thiol[7], amino[8] and pyridyl[9]. Although, this conformational change works well, they exhibit a poor over all conductance due to non-uniform binding geometries[10], metal-molecule coupling disturbing the molecular orbitals, or decoupled electron systems.[11] In order to overcome this drawbacks, direct molecule-metal coupling[12] or Au-C σ-bonds[13] are desirable. Molecule-metal coupling was used in the application of fullerenes as anchor molecules. The fullerenes bear the advantage of an enlarged π-system, but suffer from low conductance, potentially based on the weak molecule-metal coupling or intramolecular tunnelling barriers.[12] The π-system of a fullerene is not in conjugation with the core molecule because of the unfavourable geometry. Additionally, the coupling of the core molecule leads to a change in the hybridization state of one carbon atom from sp^2 to sp^3, breaking down the π-system. To elude the weak molecule-metal coupling, the anchoring molecule is bond covalently to the gold surface. Au-C σ-bonds are up to two orders of magnitude more efficient than the previously reported anchoring groups.[13] But, this single-molecule junctions don't exhibit such an amplified π-system. To combine these two systems, an Au-C σ-bond is needed, as well as an enlarged π-system within the anchor molecule, avoiding the hybridisation problem. Therefore bowl-shaped fragments of the fullerenes, so called 'bucky-bowls', were deployed. They don't suffer from a breakdown of the π-system, while being bound to the molecular core.

Here a short route towards the synthesis of a methylated bucky-ball is reported according to previous stated synthesis.[14,15] The synthesis comprises a sequence of a Wittig-Horner reaction, followed by a photocyclization. Bromination provides the starting material for a second Wittig-Horner reaction with following photocyclization. The key step of the synthesis is the cove-region closure process (CRC) facilitated by regiospecific HF elimination, promoted through activated aluminium oxide. The final product exhibits the characteristic bowl-like shape. Different reaction conditions were applied for the key step in order to increase the yield and minimize the amount of side products.

Results and Discussion

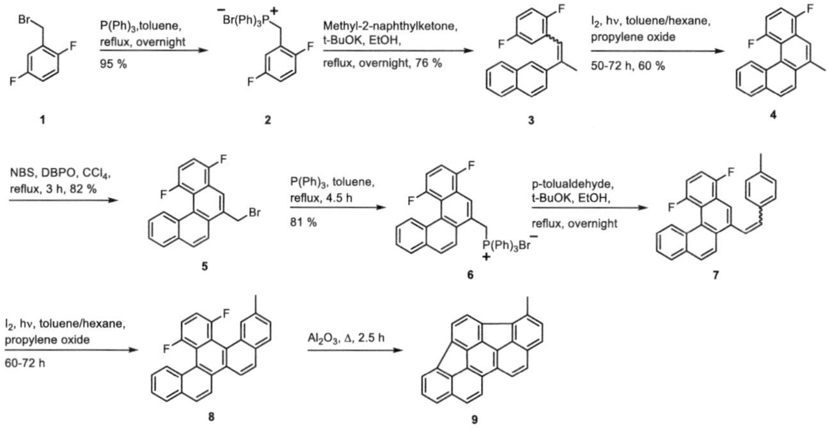

Figure 1 Overview of the reaction cascade leading to the substituted bucky-bowl (**9**)

The synthesis was started with 2-(bromomethyl)-1,4-difluorobenzene (**1**) and triphenyl phosphane to give the corresponding phosphonium salt. This was reacted in the next step with methyl-2-naphthylketone in a Wittig-Horner reaction, to give a E/Z-mixture of the benzo-stilbene (**3**) as white crystals and a yield of 72 %. Mainly the *E*-isomere was formed due to the highly stabilized phosphonium salt. The stilbene was used for the next step without separation and then converted to the substituted benzo[c]phenanthrene (**4**) using the photocyclization method as described in the literature.[14] Iodine acts as an oxidant and leads to recovery of the aromatic system. Propylene oxide binds the resulting HI. Therefore different conditions were applied in order to accelerate the reaction and to improve the yield. The fixed parameters were the benzo[c]phenanthrene derivate (1 eq) and the amount of iodine (1.05 eq). For propylene oxide a lot of different amounts used are stated in the literature. An excess (10 eq) led to a yield of 59 % and a reaction time of 60h. The lack of propylene oxide in the G-Protein gekoppelte Rezeptoren (GPCR) sind die größte Klasse von Membranprote-inen im menschlichen Genom und eine wichtige Targetklasse für derzeitige Medika-mente.

Sie folgen alle demselben strukturellen Aufbauprinzip aus sieben transmembranären α-Helices (TM), die über drei intrazelluläre und drei extrazelluläre Schleifen (ICL und ECL) miteinander verbunden sind. Bei Aktivierung durch einen endogenen oder exogenen Liganden wird eine Signalkaskade induziert, an deren erster Stelle das tri-mere G-Protein steht.

Dieses koppelt an den ICL 3, wird gespalten und migriert ent-lang der Innenseite der Membran zum Effektorsystem.

Eine interessante Gruppe sind die muskarinergen Acetylcholin Rezeptoren (mAChR), die namentlich von deren endogenen Liganden Acetylcholin und vom exogenen A-gonisten Muscarin abgeleitet sind. Neben der Expression im ZNS ist deren Präsenz im pullmonalen System von großem Interesse. Sie werden mit Atemwegserkrankun-gen, wie der chronisch obstruktive Lungenerkrankungen (COPD) und Asthma in Verbindung gebracht.

Diese Rezeptoren können in die Subklassen M1-M5 unterteilt werden. M1, M3 und M5 sind Gq/11 und M2, M4 sind Gi/o gekoppelt. Von den in menschlichen Atemwegen exprimierten Rezeptoren M1-M4, ist der M3-Rezeptor hauptsächlich auf den glatten Muskelzellen zu finden und stellt das Haupttarget für Medikamente dar. Ziel ist es, in der Therapie von COPD selektiv den Subtyp M3 zu blockieren, um eine Bronchial-verengung zu vermeiden. Simultanes Blockieren des präsynaptischen M2-Rezeptors führt zu einer abgeschwächten Wirkung[6], da dieser ein Autorezeptor ist.was around 99 %, and the yield was 60 % after the work up and purification. The E-isomere is converted into the Z isomere upon irradiation and is then cyclized. The conversion form E to Z is a slow process and the rate limiting step. Using a less stabilized phosphonium salt, the formation of the Z-isomer can be facilitated. This might lead to shorter reaction times for the photocyclization.

Compound (**4**) was brominated with NBS (yield = 82 %) to get the starting material (**5**) for the next Wittig-Horner reaction. Again, a phosphonium salt is formed, applying the conditions described above, but reducing the reaction time to 4.5 h. Using p-tolualdehyde as reagent, a mixture of E/Z methyl substituted styrylbenzo[c]phenanthrene (**7**) was obtained. The methyl group can be used later on for the coupling of the switch molecule to the bucky-bowl anchor or for extending the bucky-bowl structure. In order to get different binding patterns m-tolualdehyde and o-tolualdehyde might be used as well. The reaction exhibited high E-isomer supremacy based on the stabilized phosphonium salt. Compound (**7**) was submitted to the second photocyclization, leading to the methylbenzo[s]picene derivative (**8**). The yields of the reaction were low and the reaction time long, making this step limiting for the whole sequence. The yields might be improved and the time decreased by using a less stabilized phosphonium salt for the Wittig-Horner reaction, in order to obtain mainly the Z-isomere of compound (**7**). Using flash chromatography it is possible to separate the stilbene from the methylbenzo[s]picene.

As mentioned before, the key step of the synthesis route is the cove-region closure process (CRC) promoted by activated aluminium oxide via HF elimination, leading to methyl substituted indacenopicene (**9**). The most probable mechanism of the CRC reaction[15] is depicted in **Figure 2**.

Figure 2 Most potential mechanism of the CRC showing the key points of the process: coordination of the aluminium oxide surface, aromatic transition state, Al-F bond formation.

The formation of the new ring takes place in a concerted fashion and proceeds via a cyclic transition state. The transition state might comprises an aromatic six-membered ring, resulting in a low activation energy. The driving force of the reaction is the formation of a very strong Al-F bond. Yet there is no protocol established for a substituted indacenopicene defining suitable reaction conditions, different conditions were applied based on the literature.[15] The basic procedure was developed in the own research group and consists of the activation of γ-Al_2O_3 at 400-500 °C in a tube, whilst bubbling nitrogen through the powder. The aluminium oxide was allowed to cool down to approximately 200 °C and a small amount of the methylbenzo[s]picene derivate was added under nitrogen. The mixture was allowed to react for 60-150 min under nitrogen atmosphere. The temperature was varied between 215 and 230 °C. **Table 1** shows the different routes used to synthesise the methylindacenopicene (**9**).

m(Al_2O_3) [g]	$t_{activation}$ [min]	$T_{activation}$ [°C]	$t_{reaction}$ [min]	$T_{reaction}$ [°C]	Inert gas during activation	Route
2.00	80	490	60	230	yes	a

2.00	120	400	120	215	yes	b
2.00	120	400	120	215	no	c
1.00	120	540	120	220	yes	d
2.50	120	540	150	220	yes	e
2.00	120	500	120	220	yes	f
2.00	120	500	90	220	yes	g

Table 1 Reaction routes (a) – (g) for the CRC reactions for compound (9)

The resulting compound (9) from the different routes (a) – (g) were analysed by HPLC and the relative abundance of the desired compound was determined regarding to the peak area. The spectra showed the presence of the methylindacenopicene (9), the methylbenzo[s]picene (8) and small remnants of the stilbene (7) that was not fully separated the methylbenzo[s]picene in the previous step. Additionally the presence of a monocyclization product is possible, but was not clearly identified. The retention time of the stilbene (7) was 8.3 min and 5.9 min of the methylbenzo[s]picene (8). The methylindacenopicene (9) was observed at 13.7 min. The conditions applied in route (a) led to moderate yields (peak area = 63 %), still exhibiting the presence of the starting material. Upon increasing the activation time as well as the reaction time and decreasing the activation temperature as well as the reaction temperature the yields in (b) and (c) dropped significantly (4-7 %), regardless of the activation being carried out under nitrogen atmosphere or not. In the next step the amount of γ-Al$_2$O$_3$ was varied and the temperature was raised again. Intriguingly the yield in route (d) (60 %) was comparable to route (a), whereas route (e) just showed half the conversion (30 %). The activation temperature being around 500 °C seemed to be mandatory. Based on this finding the conditions applied in route (a) were adjusted, regarding the reaction time. Longer reaction time and dissolving the methylbenzo[s]picene in dichlormethane before administration resulted in poor conversion for route (f) (13 %). By shortening the reaction time to 90 min the conversion was raised to high levels in route (g) (85 %).

A clear dependence of the reaction time and temperature could not be determined, but the activation temperature seems to be a limiting factor for the CRC. If the activation temperature is around 500 °C, the amount of γ-Al$_2$O$_3$ has to be tailored to the reaction time. The activation time seems just to be a minor contributor, if a certain period is exceeded.

Dissolving the methylbenzo[s]picene before adding to the reaction, didn't exhibit any improvements. Based on the results of (**9g**) the conditions may be fine-tuned in order to improve the level of conversion.

Summary and Outlook

The proposed synthesis route could successfully employed. The precursor methylbenzo[s]picene (**8**) for the CRC reaction was obtained and some studies for the key reaction were carried out, applying different routes (**a**) – (**g**). The conditions for this reaction have still to be refined in order to get better conversion and reproducible results. Furthermore the Wittig-Horner reactions can be modified to increase the ratio of E/Z isomere mixture. This would lead to faster and better conversions of the stilbenes in the photocyclization. The next steps would also comprise the usage of m-tolualdehyde or o-tolualdehyde in order to generate different kind of anchor molecules.

Experimental Section

General Methods. All commercially available reagents and solvents were used without further purification, if not stated otherwise. R_f values were determined on TLC-PET sheets coated with silica gel and a fluorescent indicator (254 nm). Column chromatography was performed on silica gel Kieselgel 60. For the photocyclization a UV lamp (125 W) was used. ^{1}H NMR and ^{13}H NMR spectra were recorded on a 400.13 MHz Brucker Avance III NMR spectrometer.

HPLC analysis were performed on a Hewlett Packard Series 1100 (no gradient, 25 min runtime, pure acetonitrile).

GC-MS analysis were carried out on an Agilent Technologies 7850A GC-System (250 °C, various time of flight), connected to a Hewlett Packard HP 6890 Series Mass Spectrometer.

Synthesis of (*E/Z*)-2-(1-(2,5-difluorophenyl)prop-1-en-2-yl)naphthalene[14] (**3**)

To a solution of 2-(bromomethyl)-1,4-difluorobenzene (**1**) (10.00 g, 48.3 mmol, 1 eq) in toluene (150 mL) was added triphenylphosphane (13.30 g, 50.7 mmol, 1.05 eq). The reaction mixture was heated at reflux with strong stirring overnight. A white, crystalline solid precipitated from the solution. The reaction mixture was cooled to room temperature and then filtered. The solid was washed with toluene and petroleum ether twice and dried under reduced pressure. It yielded the phosphonium salt (**2**) (21.43 g, 45.7 mmol, 95%) as a white powder. The resulting salt was used in the next step without further purification.

The phosphonium salt (**2**) (21.43 g, 45.7 mmol, 1 eq) was dissolved in dry ethanol (200 mL) under an inert gas atmosphere (N_2). Methyl-2-naphthylketon (7.28 g, 45.7 mmol, 1 eq) was dissolved under nitrogen in 15 mL of dry ethanol and added under stirring. The colour was light grey. t-BuOK (5.13 g, 45.7 mmol, 1 eq) was dissolved in 15 mL of dry ethanol and slowly added over 5-10 minutes. The colour of the solution changed to orange. The mixture was heated to reflux and stirred overnight. The reaction was monitored with TLC. The colour changed to milky grey. The mixture was cooled to room temperature, concentrated by evaporation under reduced pressure, diluted with water and extracted with EtOAc (150 mL x 4). The combined organic phases were dried over Na_2SO_4, filtered and concentrated by evaporation under reduced pressure. The crude product was purified by flash chromatography on silica gel using a mixture of petroleum ether (PE) and EtOAc (6:1) as eluent. All fractions containing the desired product were combined and the solvent removed under reduced pressure. The E/Z mixture of the stilbene (**3**) (9.72 g, 34.7 mmol, 76%) was obtained as a white solid.

R_f (**3**) = 0.68 (PE/EE 5:1).

Synthesis of 1,4-difluoro-6-methylbenzo[c]phenanthrene[14] (4)

In a 125 W water-cooled photochemical reactor the E/Z mixture of the benzo-stilbene (**3**) (3.00 g, 10.7 mmol, 1 eq) was dissolved in toluene (30 mL) and filled up with cyclohexane. Nitrogen was bubbled through the stirred solution for around 15 min. Iodine (2.99 g, 11.8 mmol, 1.1 eq) and an excess of propylene oxide (22.5 mL, 321.2 mmol, 30 eq) were added. After irradiation for 15-20 h the colour of iodione had disappeared. The conversion of the benzo-stilbene was monitored with GC-MS. Iodine and propylene oxide were added in respective amounts, according to the amount of remaining benzo-stilbene (**3**). This procedure was repeated as soon as the colour of iodine had disappeared, until nearly full conversion of the benzo-stilbene. The total amounts of reagents added during the reaction were iodine (6.49 g, 25.6 mmol) and propylene oxide (50.0 mL, 713.8 mmol). The total time was 48 h. The reaction mixture was washed with aqueous $Na_2S_2O_3$ (100mL x 3) to remove residual traces of iodine, dried over Na_2SO_4, filtered and concentrated under reduced pressure. The crude product was purified with flash chromatography on silica gel using PE as eluent. The cyclization product (**4**) (2.10g, 7.5 mmol, 60%) was obtained as a white powder.

R_f (**4**) = 0.31 (PE). ^{1}H NMR (400.13 MHz, CDCl$_3$) δ 8.36 – 8.21 (m, 1H), 8.12 – 7.92 (m, 4H), 7.74 – 7.52 (m, 2H), 7.33 – 7.21 (m, 2H), 2.86 (s, 3H); ^{13}C NMR (100.61 MHz, CDCl$_3$) δ 155.3 (dd, J_1 = 249.51 Hz, J_2 = 3.02), 154.7 (dd, , J_1 = 244.48 Hz, J_2 = 2.01 Hz), 134.4 (dd, J_1 = 2.01 Hz, J_2 = 1.01 Hz), 132.3 (d, J = 58.35 Hz), 130.1 (d, J = 16.01), 130.0 (d, J = 3.02 Hz), 128.9, 127.3 (d, J = 1.01 Hz), 126.7, 126.4 (d, J = 1.01 Hz), 126.2 (d, J = 4.02 Hz), 125.1 (d, J = 3.02 Hz), 124.8 (dd, J_1 = 4.02 Hz , J_2 = 2.01 Hz) , 124.0 (dd, J_1 = 18.11 Hz, J_2 = 5.03 Hz), 121.8, 119.1 (dd, J_1 = 7.04 Hz, J_2 = 2.01 Hz), 111.1 (dd, J_1 = 27.16 Hz, J_2 = 10.06 Hz), 110.7 (dd, J_1 = 23.14 Hz, J_2 = 9.05 Hz), 20.6.

Synthesis of 6-(bromomethyl)-1,4-difluorobenzo[c]phenanthrene[14] (5)

The respective methylbenzo[c]phenanthrene derivate (4) (2.26 g, 8.1 mmol, 1 eq) was dissolved in CCl_4 (180 mL). N-bromosuccinimide (NBS; 1.45 g, 8.1 mmol, 1 eq) and a catalytic amount of dibenzoyl peroxide (DPBO; 5-10 mg) were added. The mixture was heated at reflux and the progress was monitored with TLC every hour. After 3h the mixture was allowed to cool down. During the reaction, the solution changed from slightly turbid to clear. NBS was filtered off and the solvent was evaporated under reduced pressure. The resulting yellow-brownish solid was sputtered by ultrasonic in PE, filtered and washed with PE (3x) to remove side products. After drying under reduced pressure, the product (5) (2.36 g, 6.6 mmol, 82%) was obtained as a white solid.

R_f (5) = 0.17 (PE). ^{1}H NMR (400.13 MHz, CDCl$_3$) δ 8.32 – 8.17 (m, 2H), 8.21 (d, J = 12.0 Hz, 1H), 8.14 (d, J = 12.0 Hz, 1H), 8.08 – 8.01 (m, 1H), 7.70 – 7.62 (m, 2H), 7.38 – 7.30 (m, 2H), 5.07 (s, 2H); ^{13}C NMR (100.61 MHz, CDCl$_3$) δ 155.2 (dd, J_1 = 250.52 Hz, J_2 = 3.02 Hz), 155.1 (dd, J_1 = 247.00 Hz, J_2 = 2.52 Hz), 133.3 (dd, J_1 = 3.02 Hz, J_2 = 1.01 Hz), 132.8 (d, J = 1.01 Hz), 130.0, 130.0, 129.9, 129.9, 129.4, 127.4 (d, J = 2.01 Hz), 126.9, 125.4 (d, J = 3.02), 123.4 (dd, J_1 = 17.10 Hz, J_2 = 5.03 Hz), 121.2, 120.8 (dd, J_1 = 7.04 Hz, J_2 = 2.01 Hz), 113.0 (dd, J_1 = 27.67 Hz, J_2 = 9.56 Hz), 111.3 (dd, J_1 = 19.12 Hz, J_2 = 4.53 Hz), 31.8 (1 signal was not observed due to overlapping).

Synthesis of (E/Z)-1,4-difluoro-6-(4-methylstyryl)benzo[c]phenanthrene[14] (7)

To a solution of the respective bromide (5) (2.26 g, 6.3 mmol, 1 eq) in toluene (100 mL) was added triphenylphosphane (1.74 g, 6.6 mmol, 1.05 eq). The reaction mixture was heated at reflux with strong stirring for 4.5 h. A white crystalline solid precipitated from the solution. The reaction mixture was cooled to room temperature and then filtered. The solid was washed twice with toluene and petroleum ether and dried under reduced pressure. The phosphonium salt (6) (3.16 g, 5.1 mmol, 81%) was obtained as a slightly yellow powder. The resulting salt was used in the next step without further purification.

The phosphonium salt (6) (3.16 g, 5.1 mmol, 1 eq) was dissolved in dry ethanol (125 mL) under an inert gas atmosphere (N_2). The freshly destilled p-tolualdehyde (0.6 mL, 5.1 mmol, 1 eq) was added with stirring to the solution. The colour was light grey. t-BuOK (0.57 g, 5.1 mmol, 1 eq) was dissolved in a small amount of dry ethanol and slowly added over 5-10 minutes. The colour of the solution changed to orange. The mixture was heated to reflux and stirred overnight. The reaction was monitored with TLC. The colour changed to milky grey. The mixture was cooled to room temperature, concentrated by evaporation under reduced pressure, diluted with water and extracted with EtOAc (150 mL x 4). The combined organic phases were dried over Na_2SO_4, filtered and concentrated by evaporation under reduced pressure. The crude product was purified by flash chromatography on silica gel using PE as eluent. The fractions containing either the E- and Z-isomere were collected separately. The solvent was removed under reduced pressure. The E-isomer was obtained as a clear, yellow and viscous liquid. The Z-isomer was obtained as a yellowish powder. The yield was not determined.

R_f (Z-7) = 0.20 (PE), R_f (E-7) = 0.28 (PE).

Synthesis of 13,16-difluoro-2-methylbenzo[s]picene[14] (**8**)

In a 125 W water-cooled photochemical reactor the *E/Z* mixture of the benzo-stilbene (**7**) (2.65 g, 7.0 mmol, 1 eq) was dissolved in toluene (30 mL) and filled up with cyclohexane. Nitrogen was bubbled through the stirred solution for around 15 min. Iodine (1.95 g, 7.7 mmol, 1.1 eq) and an excess of propylene oxide (14.6 mL, 209.0 mmol, 30 eq) were added. After irradiation for 15-20 h the colour of iodione had disappeared. The conversion of the benzo-stilbene was monitored with GC-MS. Iodine and propylene oxide were added in respective amounts, according to the amount of remaining benzo-stilbene (**7**). This procedure was repeated as soon as the colour of iodine had disappeared, until nearly full conversion of the benzo-stilbene. The total amounts of reagents, added during the reaction, were iodine (3.95 g, 155.6 mmol) and propylene oxide (29.6 mL, 423.0 mmol). The reaction mixture was washed with aqueous $Na_2S_2O_3$ (100mL x 3) to remove residual traces of iodine, dried over Na_2SO_4, filtered and concentrated under reduced pressure. The crude product was purified by flash chromatography on silica gel using a mixture of PE/EtOAc (20:1) as eluent. The cyclization product (**8**) was obtained as a white solid.

R_f (**8**) = 0.12 (PE). ^{1}H NMR (400.13 MHz, CDCl$_3$) δ 8.53 (d, J = 8.9 Hz, 1H), 8.47 (d, J = 8.0 Hz, 1H), 8.26 – 8.14 (m, 1H), 8.10 (d, J = 8.8 Hz, 1H), 8.07 (d, J = 9.0 Hz, 1H), 8.03 (d, J = 4.0 Hz, 1H), 7.87 (d, J = 8.2 Hz, 1H), 7.65 – 7.52 (m, 2H), 7.45 – 7.33 (m, 2H), 7.13 (d, J = 8.0 Hz, 1H), 6.75 (d, J = 7.8 Hz, 1H), 2.60 (s, 3H); ^{13}C NMR (100.61 MHz, CDCl$_3$) δ 155.3 (dd, J$_1$ = 250.52 Hz, J$_2$ = 3.02 Hz), 136.5, 135.1 (d, J = 3.02 Hz), 132.4 (dd, J$_1$ = 3.02 Hz, J$_2$ = 1.01 Hz), 130.7-130.3 (m, 4C), 129.6 (dd, J$_1$ = 46.28 Hz, J$_2$ = 14.09 Hz) 129.0 (dd, J$_1$ = 47.75 Hz, J$_2$ = 16.60 Hz), , 128.4, 128.0, 127.3 (d, J = 1.01 Hz), 127.1, 126.3, 126.2, 125.3 (d, J = 3.02 Hz), 125.0, 121.9, 120.0, 119.0, 114.4 (dd, J$_1$ = 27.16 Hz, J$_2$ = 10.06 Hz), 114.2 (dd, J$_1$ = 28.17 Hz, J$_2$ = 10.06 Hz), 22.0 (two signals were not observed due to overlapping).

Synthesis of 3-methyl-*as*-indaceno[3,2,1,8,7,6-*pqrstuv*]picene[15] (**9**)

Aluminiumoxide (1.00-2.50 g) was activated by annealing (400-540 °C) in a glas tube for 80-120 min under inert gas atmosphere. The mixture was allowed to cool down to reaction temperature (215-230 °C) and a small amount (~10 mg) of compound (**8**) was added. The mixture was heated for 60-150 min under inert gas atmosphere. Afterwards the mixture was allowed to cool down to room temperature. It was extracted with hot toluene (30 mL x 3). Toluene was removed under reduced pressure. The product (**9**) was obtained as a yellow solid.

R_f (**9**) = 0.48 (PE/DCM 3:1). ^{1}H NMR (400.13 MHz, CDCl$_3$) δ 8.11 (d, J = 8.8 Hz, 1H), 8.06 (d, J = 8.7 Hz, 1H), 7.77 (t, J = 8.9 Hz, 2H), 7.71 (d, J = 7.0 Hz, 2H), 7.67 (d, J = 7.3 Hz, 1H), 7.64 (d, J = 8.2 Hz, 1H), 7.55 (d, J = 8.2 Hz, 1H), 7.39 (dd, J$_1$ = 8.1 Hz, J$_2$ = 6.9 Hz, 1H) 7.19 (d, J = 8.3 Hz, 1H), 2.75 (s, 3H).

^{13}C NMR (100.61 MHz, CDCl$_3$) δ 139.3, 139.1, 138.6, 138.5, 138.1, 138.0, 137.7, 137.1, 137.0, 136.9, 136.0, 132.0, 130.2, 129.2, 128.8, 127.6, 127.1, 127.0, 126.8, 126.7, 125.8, 125.6, 124.1, 123.5, 123.2, 20.3 (one signal was not observed).

Literature

1. A. Aviram and M. A. Ratner, Chem. Phys. Lett. **29**, 277 (1974)
2. C. Joachim *et al.*, Nature **408**, 541 (2000)
3. S. V. Aradhya and L. Venkataraman, Nature Nanotechnol. **8**, 399 (2013)
4. D. Dulic *et al.*, Phys. Rev. Lett. **91**, 207402 (2003)
5. T. Sendler *et al.*, Adv. Sci. **2**, 1500017 (2015)
6. Y. Kim *et al.*, Nano Lett. **12**, 3736 (2012)
7. F. Chen *et al.*, J. Am. Chem. Soc. **128**, 15874 (2006)
8. M. S. Hybertsen *et al.*, **20**, 374115 (2008)
9. B. Xu, *et al.*, **301**, 1221 (2003)
10. J. S. Kristensen *et al.*, J. Phys.: Condens. Matter **20**, 374101 (2008)
11. J. C. Cuevas, E. Scheer, World Scientific Series in Nanoscience and Nanotechnology; World Scientific: Singapore, 2010
12. E. Lörtscher *et al.*, Small **9**, 209 (2013)
13. W. Hong, *et al.*, J. Am. Chem. Soc. **134**, 19425 (2012
14. K. Y. Amsharov *et al.*, Eur. J. Org. Chem. **36**, 6328 (2009)
15. K. Y. Amsharov *et al.*, Angew. Chem. **124**, 4672 (2012)

BEI GRIN MACHT SICH IHR WISSEN BEZAHLT

- Wir veröffentlichen Ihre Hausarbeit, Bachelor- und Masterarbeit

- Ihr eigenes eBook und Buch - weltweit in allen wichtigen Shops

- Verdienen Sie an jedem Verkauf

Jetzt bei www.GRIN.com hochladen und kostenlos publizieren